A Picture Book For Our Universe

Written & Illustrated by Mack Spellman

Keir McCormack H. Spellman
December 31, 2020

to the ones I love...

Once upon a time there was nothing, and then there was something. Our universe is a beautiful, mysterious, adventurous place, and by exploring the history of scientific discoveries we can better understand the magic of infinity...

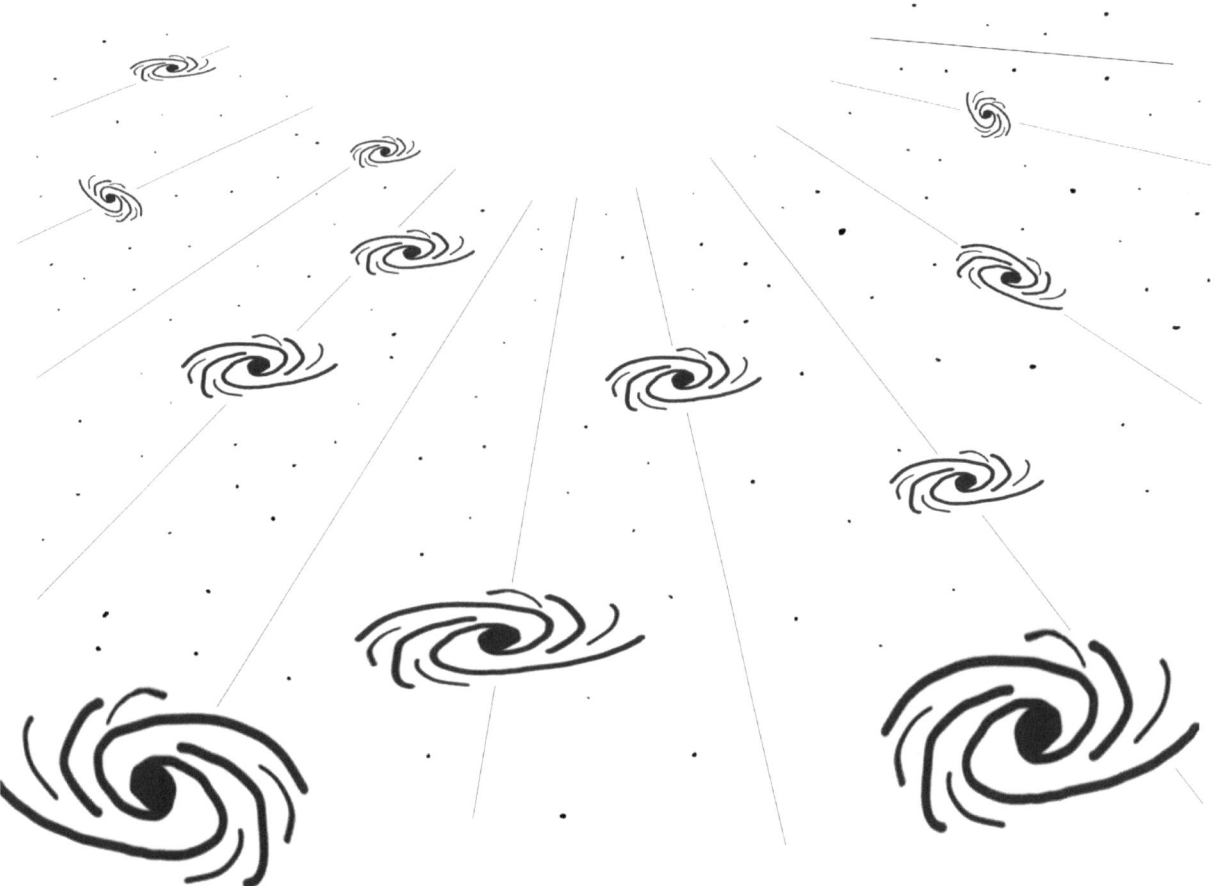

The earliest known system of measures originated in Egypt and Mesopotamia around 3000 BCE. The cubit was the length of the forearm from the elbow to the tip of the middle finger, a grain of barley or wheat was used as a unit of mass, and by tracking the movement of the stars the Egyptians developed a 365 day solar calendar. Yet, these unit measurements would vary, and a system of agreed value was needed in order to further progress...

It wasn't until the 1400's when the first mechanical clocks were able to display fractions of the day more precisely. By using energy from a wound coil to spin a series of wheels, a sawtooth-shaped crown wheel prevents the energy from escaping all at once. This allowed days to be subdivided into minutes and seconds advancing at regular intervals, or "ticks"...

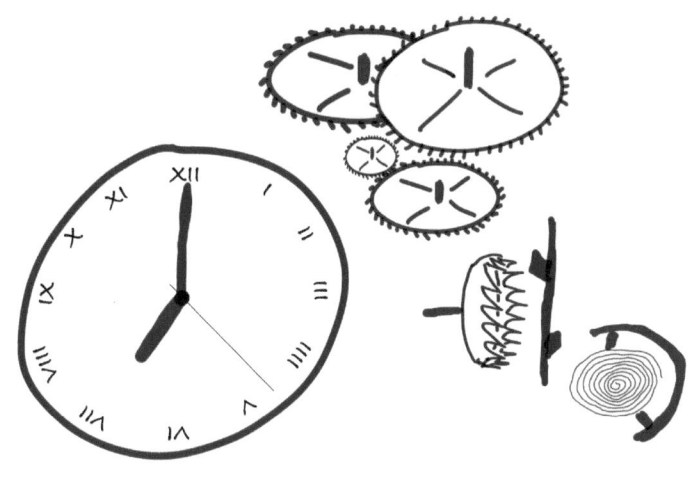

1	13	25	37	49
2	14	26	38	50
3	15	27	39	51
4	16	28	40	52
5	17	29	41	53
6	18	30	42	54
7	19	31	43	55
8	20	32	44	56
9	21	33	45	57
10	22	34	46	58
11	23	35	47	59
12	24	36	48	60

During the 1600's, a man named Isaac Newton had studied why objects with mass fall towards Earth, and why celestial bodies seemed to attract towards one another. He found that as objects fall they will accelerate moving faster and faster until they hit the ground, discovering a mysterious force called gravity...

By the French Revolution, pendulums were used to determine unit measurements more accurately, providing balance between length, mass, time, and gravity. The second was officially defined as "$1/86,400$ of a mean solar day," the meter was set as "the length based on a pendulum with a half-period of one second," and the kilogram was set using the weight of water as "the mass of a cubic decimeter"...

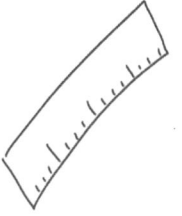

In the beginning of the 20th century, Albert Einstein would redefine our entire understanding of the physical world. By predicting how starlight would bend around the sun during an eclipse, he determined that gravity was not an attractive force, but rather the result of mass actually warping space and time. This in turn proved that Light travels at a constant speed, relative to everything...

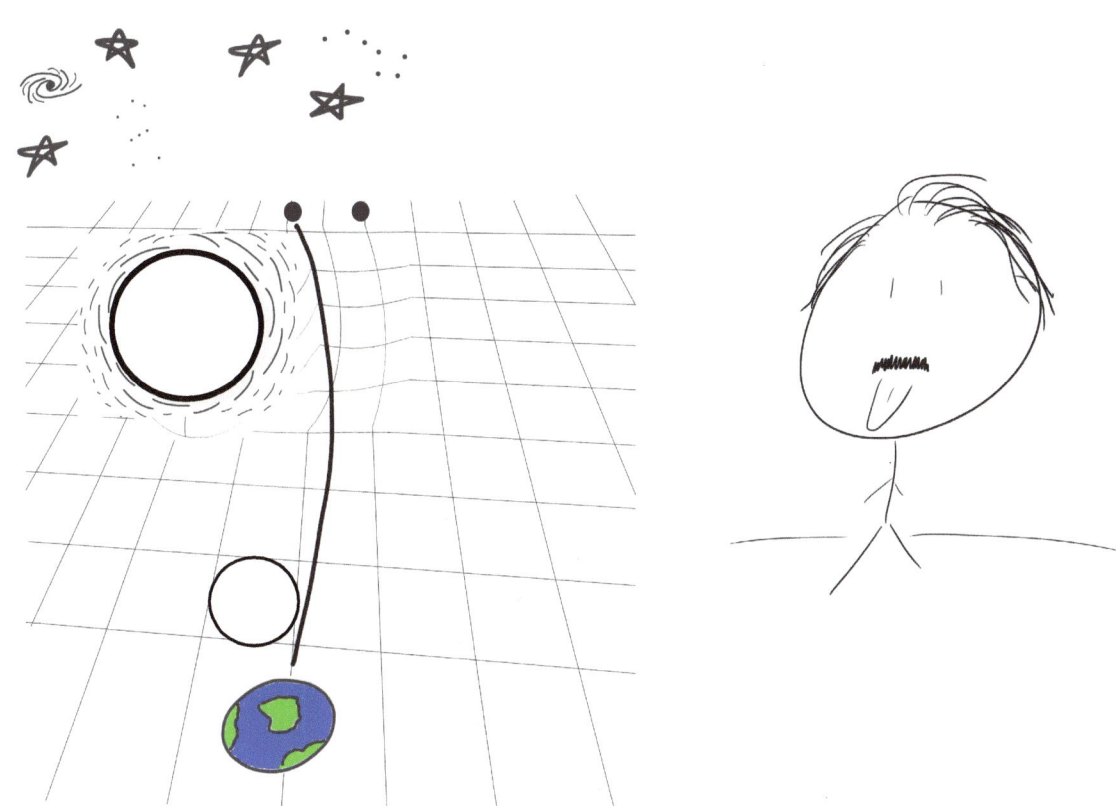

The Speed of Light is the fastest thing in our universe, and it has amazed many of history's greatest minds. Back in 1676, Danish astronomer Ole Rømer discovered Light does not move instantaneously. He noticed that during the time of year when Earth was closer in orbit to Jupiter, one of it's moons would eclipse 11 minutes faster than he expected. Six months later when Earth was farther away in orbit from Jupiter, the same moon would eclipse 11 minutes later than his calculations. This led Rømer to determine that it would therefore take roughly 22 minutes for Light to travel the diameter of Earth's orbit...

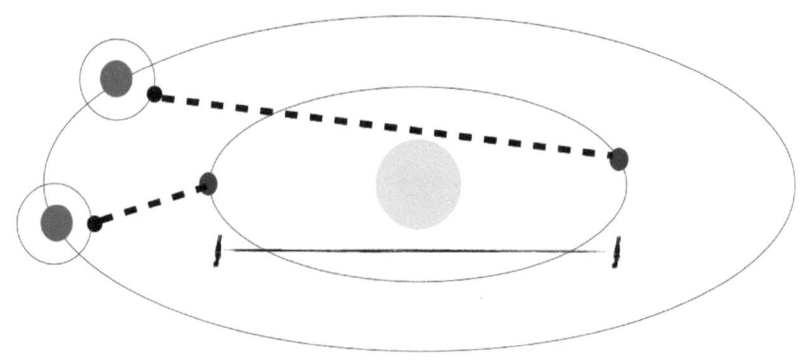

During the 1800's, scientists began experimenting with electric current and magnetism. Through finding that the waves and frequency between these two forces matched the Speed of Light, James Clerk Maxwell uncovered a profound relationship to be called "electromagnetism." This discovery led to our understanding of the electromagnetic spectrum, including all the colors we see...

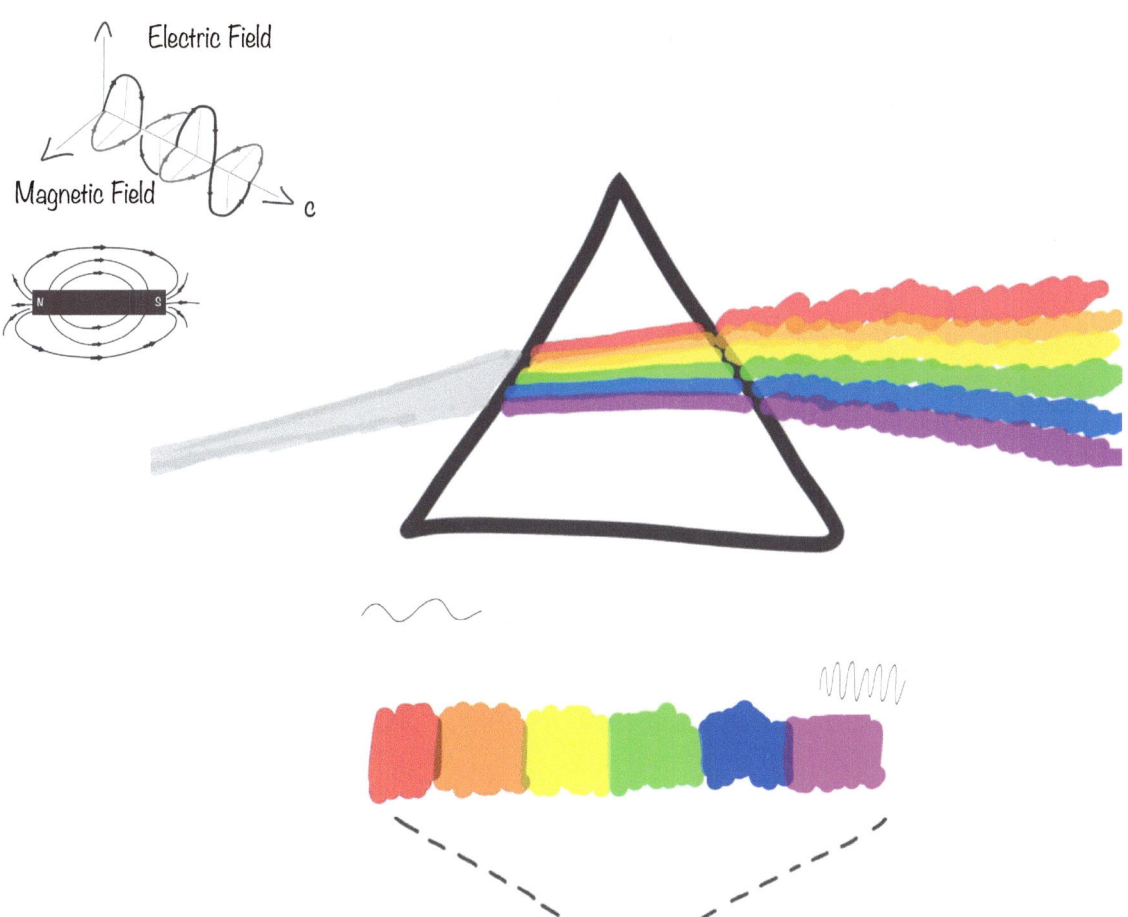

Radio | Microwaves | Infrared | **Visible Light** | Ultraviolet | X-rays | Gamma

During his miracle year of 1905, Einstein proposed that all observers regardless of their state of motion, will measure the same Speed of Light. This means that as objects move faster approaching a universal speed limit, time will pass more slowly and the object will become smaller in length. If a baseball is thrown inside of a moving train, it will appear at different velocities to an observer on the train as opposed to watching from the outside platform. Yet, a beam of Light on a moving train would be observed at the exact same velocity for both observers on and off the train. In addition to this stunning realization, Einstein put forth a relationship between energy, mass, and the Speed of Light that would revolutionize modern physics, $E=mc^2$...

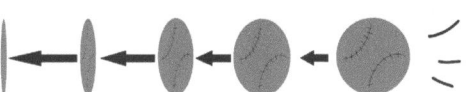
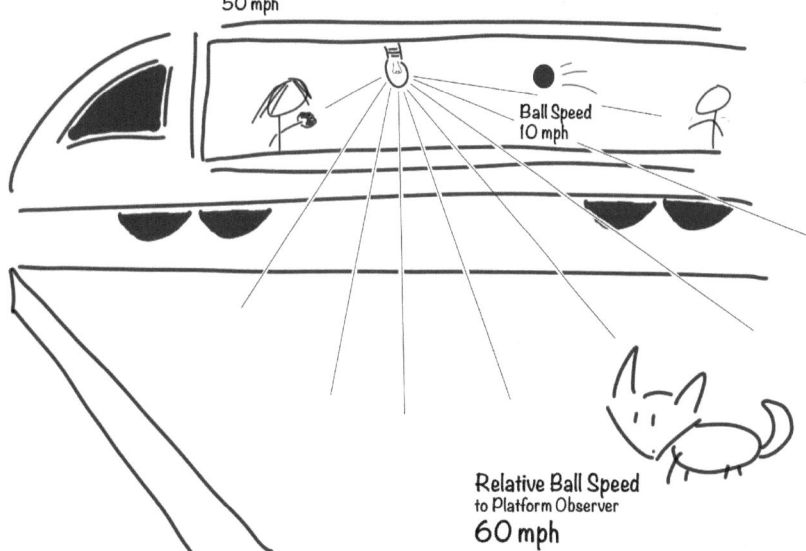

Energy exists all around us as the ability to do work. Scientific discovery has revealed that energy is interconnected and has many interchangeable forms. Every object will remain at rest or in uniform motion unless compelled to change by an external force, the force of an object changing in speed is equal to it's mass multiplied by it's acceleration, and for every action there is an equal and opposite reaction...

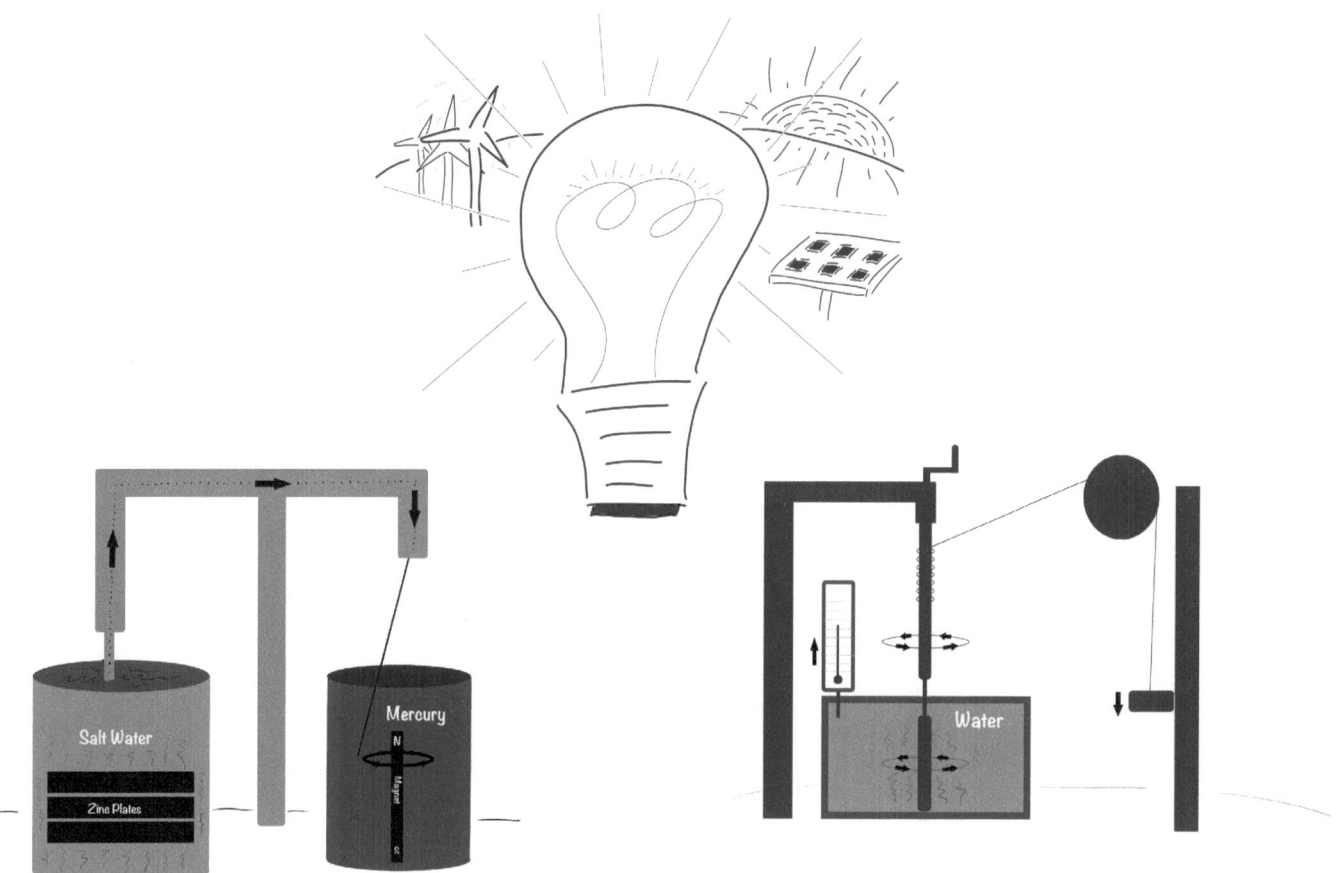

Knowledge of Light and energy has helped us understand the smallest scale of our universe through the structure of atoms. In the early 1900's, Max Planck had theorized that light must behave as individual packets of energy, like water flowing in droplets rather than a continuous stream. This notion led to Einstein's Photoelectric Effect, displaying that violet light would be capable of impacting electric particles, but red light with a longer wavelength and lower frequency would not. Soon after, Niels Bohr would explain that this incremental amount of light energy provided stability to the orbits within atoms...

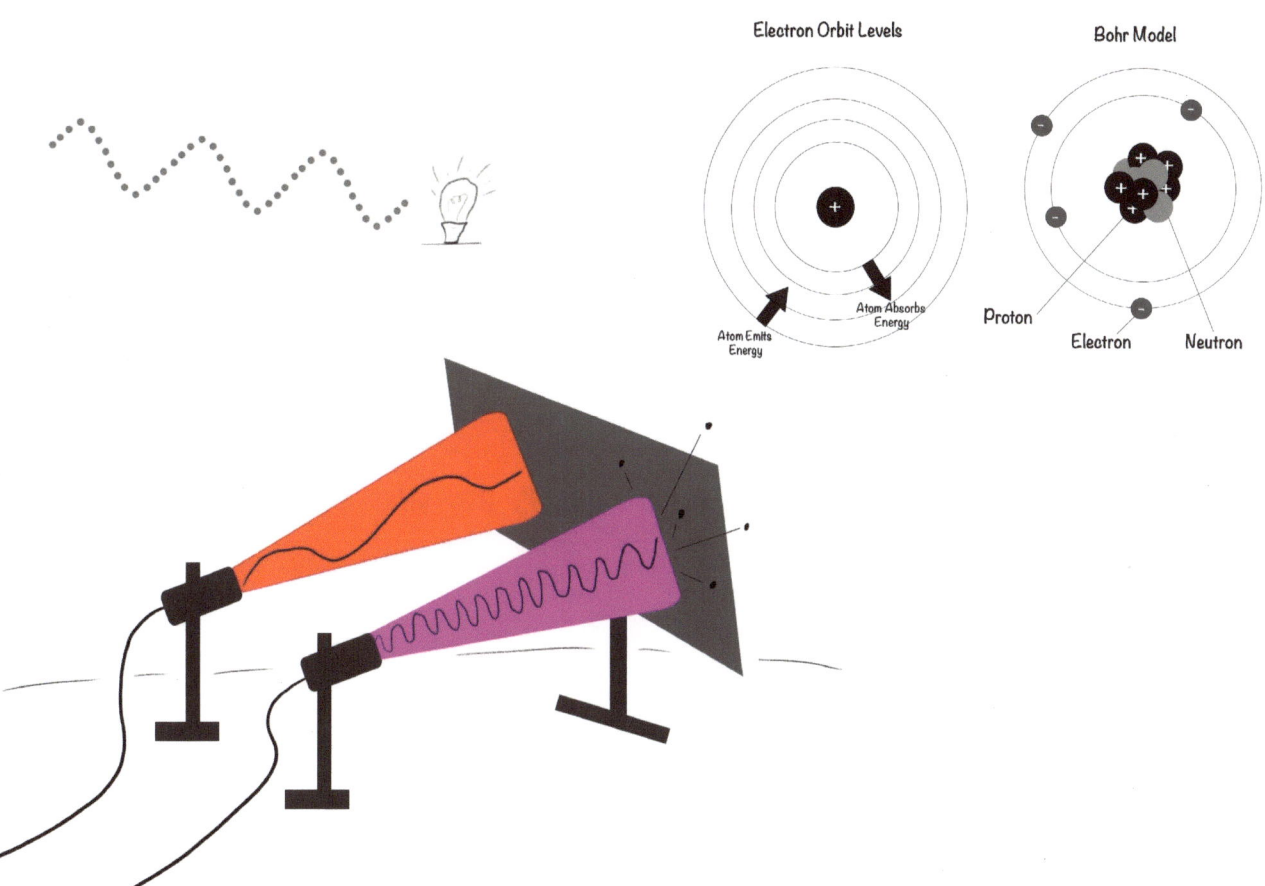

On the largest scale, Edwin Hubble used his telescope to discover that our Milky Way Galaxy is just one of many. He also used the color spectrum to determine our universe is expanding. In every direction Light observed from distant galaxies stretches longer and less frequent towards the red end of the spectrum, meaning they are all moving away from Earth. By calculating this expansion rate, and by rewinding to a point of beginning, it indicates our universe began roughly 13.8 billion years ago in a cosmic event known as the Big Bang...

Today we seek to unite the science of the very big, with the very small, searching to find a Theory of Everything. In order to better understand the balance and harmony of our universe, we must embrace the infinite nature of π...

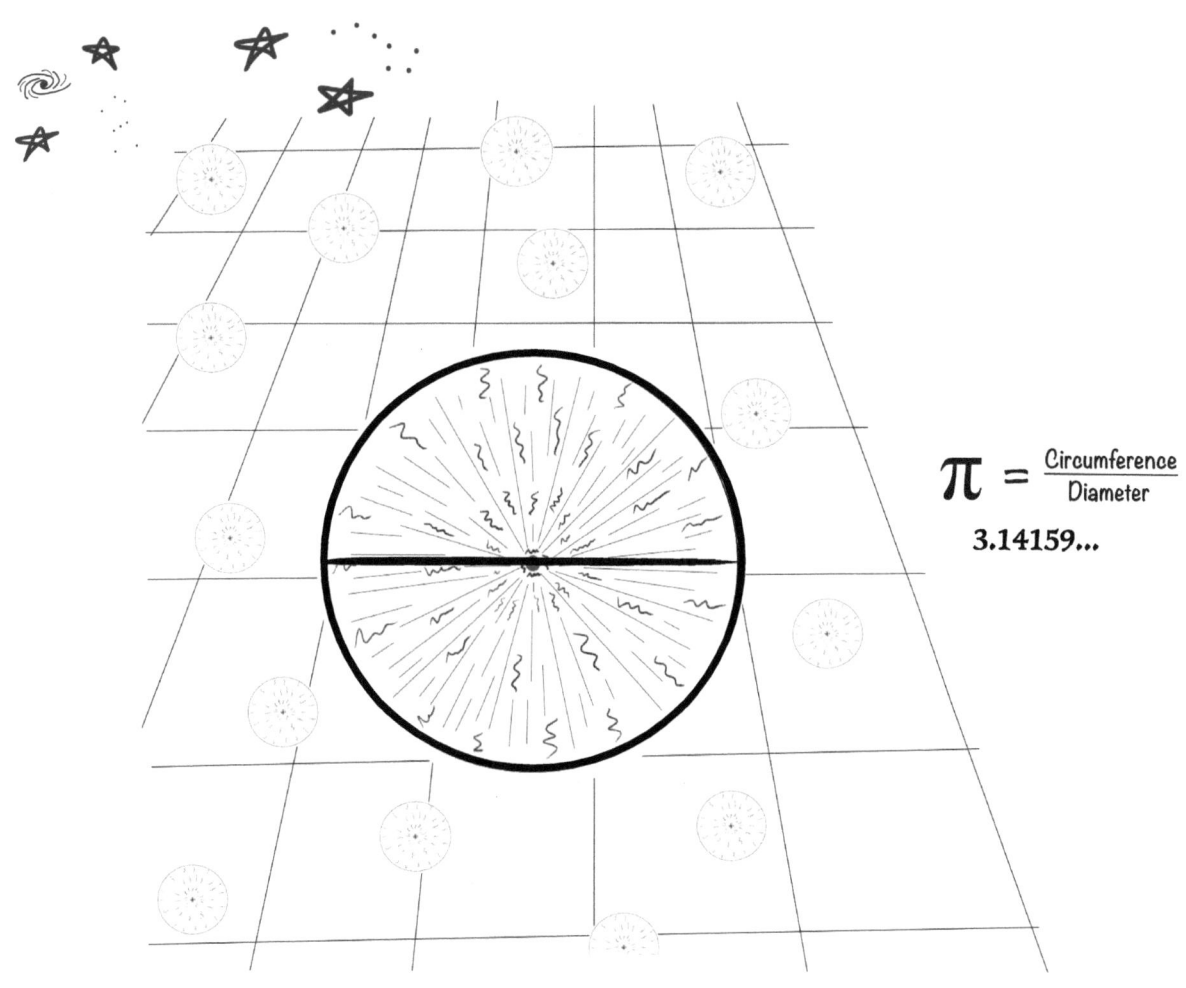

π is a transcendent mathematical constant that represents the ratio of a circle's circumference to its diameter. Approximated as 3.14159..., its decimal never ends or settles into a permanent repeating pattern. "e" is also a transcendent mathematical constant, representing a curve as the sum of an infinite series. Approximated as 2.71828..., its decimal too goes on forever. Often considered the most beautiful equation, there is a fundamental relationship between 1, 0, π, e, and the imaginary number "i" used in describing a lateral set of numbers. First discovered by Leonhard Euler, the magic of our universe may best be described through "Euler's Identity"...

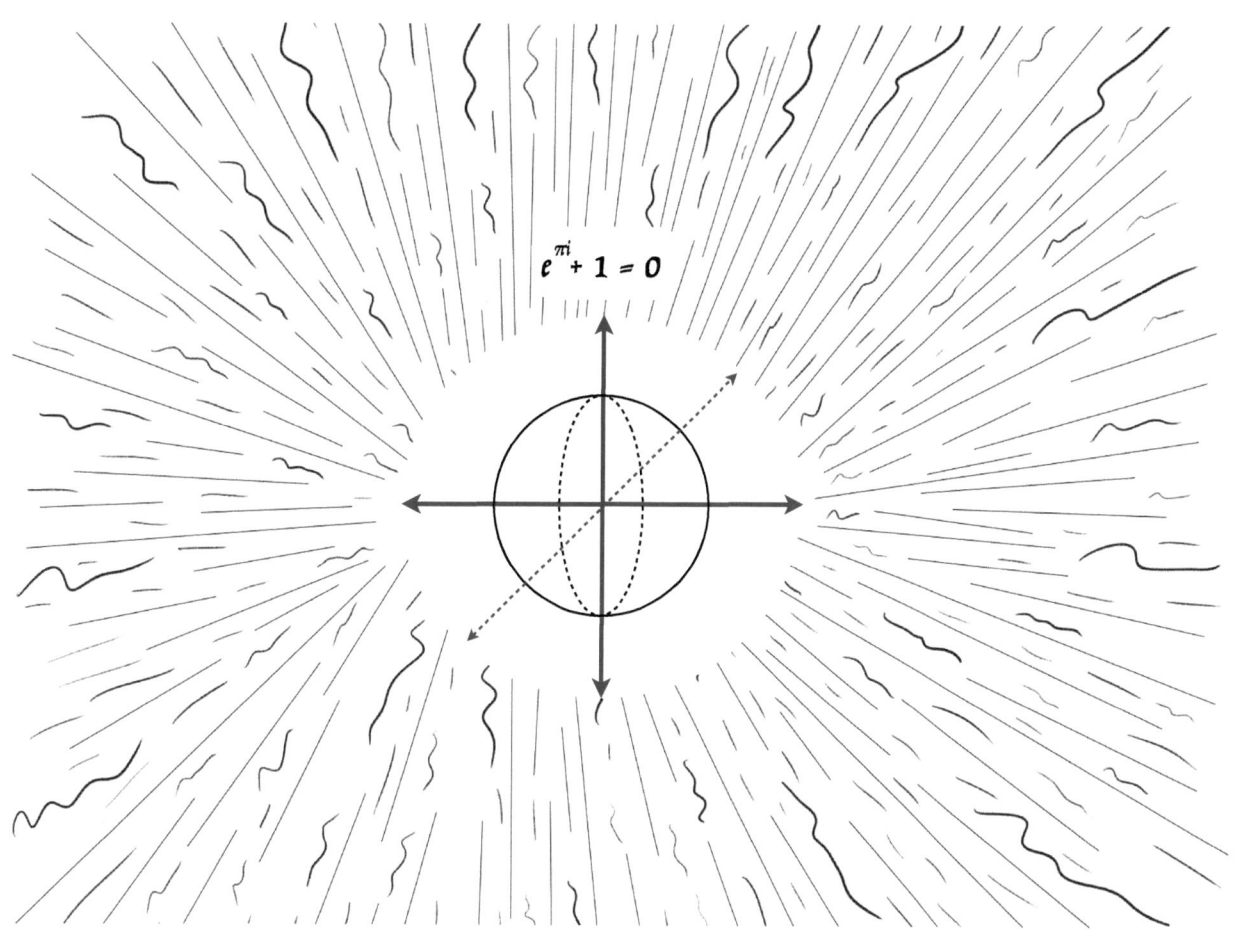

Why is standard gravity (g) 9.8 m/s²? What is the significance of the gravitational constant (G) and Planck's constant (h)? My quest began with curiosity, why is the Speed of Light (c) the fastest thing in our universe? By using a pendulum to simultaneously determine unit measurements for length, mass, and time, you will find that our universal constants share a relationship between transcendent numbers and the measurements of Earth...

$g = \pi^2$ m/s²

By using a pendulum to simultaneously determine unit measurements, the acceleration due to Earth's Gravity must equal a value of π^2 meters per second squared.

Current value of the Acceleration due to Standard Gravity on Earth (g):
9.80665 m/s²

Equation for a Half-Period of a Pendulum:
$T = \pi \sqrt{(L/g)}$

Insert "1 second" as unit for time:
$1 \text{ second} = \pi \sqrt{(L/g)}$

Insert "1 meter" as unit for length:
$1 \text{ second} = \pi \sqrt{(1 \text{ meter}/g)}$

Solve for g:
$g = \pi^2$ m/s²

$G = (\pi^2)(r^2)$ m³/earths·s²

By using the "Earth" as a unit for mass rather than the kilogram, the Gravitational Constant can be seen as a measure to Earth's core.

Current value of Newton's Gravitational Constant (G):
6.67408×10⁻¹¹ m³/kg·s²

Mass of Earth in kilograms:
5.972×10²⁴ kg

Newton's Law of Universal Gravitation:
Force = G[(m × m)/r²]

Insert "Newton's Second Law" for Force:
(m)(a) = G[(m × m)/r²]

Insert "π² m/s²" for the established Acceleration due to Earth's Gravity:
(m)(π²) = G[(m × m)/r²]

Insert "Earth" as unit for mass:
(earth)(π²) = G[(earth × earth)/r²]

Solve for G:
$G = (\pi^2)(r^2)$ m³/earths·s²

$h = (\pi/e)/c$ earths·meters²/s

The smallest observable limit of Planck's Constant can be represented by a relationship between π, e, the Speed of Light, and the mass of Earth.

Current value of Planck's Constant (h):
6.62607015×10⁻³⁴ kg·m²/s

Planck-Einstein Relation:
E=hf

Insert "Einstein's Mass-Energy Equivalence" for Energy:
mc²=hf

Insert "Maxwell's Electromagnetic relationship" for Frequency:
mc²=h(c/λ)
h=cλm

Insert "Earth" as unit for mass:
h=cλ(earth)

Insert " (π/e)/c² " for wavelength λ:
h = (c) [(π/e)/ c²] (earth)

Solve for h:
h = (π/e) / c earths·meters²/s

$c = e^\pi$ earth-diameters/s

By using "Earth's Diameter" as a unit for length, the Speed of Light can be expressed through Euler's Identity.

Current value of the Speed of Light (c):
299,792,458 m/s

Diameter of Earth:
12,740,000 m

Euler's Identity:
$e^{\pi i} + 1 = 0$

Solve for c:
$c = (e^\pi)$ earth-diameters / s

Do unto others as you would have them do unto you.

www.ingramcontent.com/pod-product-compliance
Lightning Source LLC
Chambersburg PA
CBHW042323250526
R18347300001B/R183473PG45473CBX00019B/13